Nna motivated me to become an Senior Systems Engineer (IT Specialist) and Travel around Africa
By Tanehesi The Restorer

©SameTreeDifferent Branch Publishing

Same Tree Different Branch Publishing

Copyright 2022 by Kofi Piesie Research Team
Copyright © 202s by Tanehesi The Restorer

All right reserved. No part of this book may be reproduced or transmitted in any form or by any means, electronic or mechanical, including photocopying, recording, or by any information storage and retrieval systems without the written permission of the publisher. Printed in the United States of America

Auto Industry
Reducing the dependency on Carbon-based fuels
Enhancing Safety
Developing Solar Powered Spacecraft for Interplanetary Travel
1 Hour flights to any location on Geb
Driving from one place to another in a maximum of 1hour

Table of Contents

Dedication

Chapter I - Early Years on the Farm
7

Chapter II - Developing a sense of Purpose   15

Chapter III - The influences of Education on shaping my existence 21

Chapter IV - Learning the fundamentals of (Pan) Afrakanisms at Central State and Howard
28

Bibliography

DEDICATION

In honor of Iya, this book is dedicated to Nna, my arakunrin Chris, my omo Tedros, and my immediate ebi (Isaiah, Saba, Rahel, and iyawo Zebib) and extended awon ebi like bro Garfield Reid, Ankh West, Kofi Piesie, Ini-Herit Shawn Khalfani, and the Mossi Warrior Clan, the Daggar Squad, Pseudo Killas, Science with Shawn, SaRa Suten Seti, Sheena Lynne, Kecia Jones, Thurston Hargrove, John Pitts and all my supporters on social media as well as all of those who seek truth, wisdom, and understanding………. Remember that…….Tomorrow……..is not a given……..but yesterday was…….and today…..is being given………..

Those AfRaKaNs who walked the plank
Oh, how I remember that day
When those AfRaKaNs walked the plank
Oh, how I remember Goree
When those AfRaKaNs walked the plank
So many drowned in the Great Ocean
Are those AfRaKaNs who walked the plank
So many kept their devotion
Are those AfRaKaNs who walked the plank
Now many live in the States
as those AfRaKaNs who walked the plank
Must never make a mistake and forget

Those AfRaKaNs who walked the plank

CHAPTER I:
EARLY YEARS ON THE FARM

It's a nice day and reminds me of rising early in the morning on our farm in Madison, Alabama, and feeding the chickens, pigs, and other animals.

With my awon egbon, we'd run around the farm pretending to be cowboys and Indians, where usually I would be the rebellious Indian. Iya Iya Agba was very wise and very good with accounting and managing the store we had on our farm, and Iya would also assist by helping with the accounting.

Iya would also tell me stories about my heritage and remind me that Nna (Omowale) was in AfRaKa and that we are also AfRaKaN. This would also give me a sense of purpose and inspire me to want to learn as much as possible and consider the possibility of relocating back to AfRAKa permanently with Nna. One of the things that I appreciate about growing up on a farm is that it taught me all the fundamental principles needed to be a productive and successful eniyan.

I acquired the essential skills of reasoning and logic as well as the pure basics of mathematics by keeping up with the livestock and all the vegetables, grains, and fruits; which my ebi was able to produce and sell to the community in Madison, Al.

Even before being enrolled in any educational system, my ebi taught me how to assist in the management of our farm and as an odo, my primary responsibility was managing a count of all our livestock along with selling items in the store we had on the farm. From this, I could comprehend the core fundamentals of logic and math, which I would understand more as I matured.

I also learned entrepreneurship fundamentals, which would later become the catalyst that motivated me to start my own business. This also served as a source of motivation for others, including my awon Anti and my awon Aburo, which I and my awon egbon also inherited. The principles of entrepreneurship have always been an essential component of my ebi's perspective that shaped our sense of identity.

Like many other eniyan here in the capital of HK, my ebi found ways to survive under HK oppression of the descendants of AfRaKaNs primarily but also eniyan in general in a "race," "class," and "gender" based system of control.

One of my goals is to restore traditional AfRaKAN consciousness to AfRaKaNs and return HK consciousness to HK. In this way, we allow existence to escape the "white" trap imposed on us by the HK elitists.

Here is my Nna Agba George……..this Oba served this country well…………..in military service…….then came home to  harsh conditions……..just like many awon arakunrin continue to do………but still found the capacity… to remain a pillar of strength… just like many proud members of our community.

On our farm in Alabama, we lived an AfRaKaN life in an American context which I now understand was a way of preserving our AfRaKaN heritage. When I was about 5, Nna returned from Abyssinia and came to visit. I was excited and ran to greet him when Iya said, "there's your Nna"! As I reflect on this now, it is clear that within the consciousness of the AfRaKaN exists the dichotomy of two perspectives: one that is AfRaKaN and one that is Caucasian. Subliminally, Nna understood this dichotomy and was trying to tell me that he was AfRaKaN without saying AfRaKaN. It took me some time to understand this, a condition that runs heavily in the AfRaKaN in America community.

My early years on the farm owned by my ebi were nice. We had more than 333 acres of land, which members of our ebi lived on. Our farm produced fruits and vegetables, cattle, pigs, and chickens. Nna's ebi also lived on land owned by Iya's ebi. So we were very close, and my uncles always teased me when I was young. But it was all in fun, as odo would do.

As an odo, you don't understand the dynamics of living under occupation and the complications that come with it. But Nna completely understood this dichotomy and inspired me to study AfRaKa. When I look back on my consciousness, what Nna told me was equivalent to what Iya told me, which was to honor my Awon Iya/Nna nla and defend my ebi. This advice gave me a foundation to stand on and a source of guidance. With this inspiration, my young existence was propelled by the possibilities that AfRAKa offered, and I pursued education and employment opportunities that would further promote my AfRaKaN goals.

To this extent, I obtained degrees in Math, Philosophy, Computer Science, History, and various IT Certifications and training, including MCSE, AWS, Linux and Android, IOS, and others.

With all my Education and training, I was able to become an Information Technology Specialist and pursued employment opportunities with major international companies and organizations like USAID,

UNECA, Leland, IBM, ACI Worldwide, and Global Payments, in addition to running my own production company (Mill studios) as well. I worked with younger artists and activists for several years and produced many music videos and TV shows.

Some of my favorite artists to work with were Lakisha C. Brooks, Simone Butcher, Jay Ohh, Bravo, Skinny, and others from my West GA ebi.

CHAPTER II:
Developing a sense of Purpose

My perspective changed significantly with the return of Nna from Kanyo Station in my odo. Although we didn't spend much time together, I discovered a source of motivation and began learning as much as possible about AfRaKa. When I was 5, I would tell my awon egbon that I was AfRaKaN and considered myself Ethiopian and opposed to Nubian, which was inaccurate because Nubian is AfRaKaN and Ethiopian is Greek. But these are not the same, and in fact, these two people were historical enemies. However, I studied both carefully to the extent of learning their languages and culture.

Furthermore, Iyawo is Hamasan and a descendant of the Heka Kasu. So, I don't hate them at all. But to understand the current threat, we must realize perspectives and their origins. The reluctance to accept the Neter of balance could produce an extinction-level event, which I am trying to avoid. I think Nna discovered this as well but remained silent. Although we didn't spend much time together, that was a rich source of motivation that inspired me to become a strong Okunrin/Hem Neter.

Following Nna's and Dr. Joseph E. Lewis' advice, I studied the greats, including John Jackson, Dr. Chancellor Williams, Dr. Yosef Ben- Yochannan, Dr. Cheikh Anta Diop, DR. Ivan Van Sertima,m Dr. Kwame Nkrumah and Dr. John Henrik Clarke. Due to this extent, I was able to complete everything in a Ph.D. program but the dissertation from Howard because of illness.When I fell, I realized that my obi provided me with the best advice and that I had been living the principles of aye that they taught me throughout my early existence. I used my obi as guidance to continue forward in honor and aye. Still, nothing compares to the experience of health struggles and the battles Iyawo and my ebi fought against the torturous insurance companies.

At the time of this writing, Nna has completed 84 orbits, and based on an oath I gave to Iya; I am here to defend him as the Masculine Hem Neter that Nna is. I say this because, in my opinion, it took a lot of courage, strength, and honor for Nna to admit this in a society built on a caucasian existential perspective. In this regard, it would be nice to return to Ife with Nna and my ebi to experience Nigeria and the remnants of Ancient Nubia. It would be nice to see Kandakes restored to their rightful place as the awon iya of existence. The one who creates ebi deserves the honor and credit for hemetu contributions. In this regard, I will attempt to fulfill my obi's instructions.

This will be hard for many to understand, but my upbringing, research, and DNA allowed me to realize that I am existentially a descendant of Ancient Nubia, i.e., AfRAKaN. What Dr. Joseph E. Lewis and other great AFRaKaN scholars taught me was the extent to which I had the responsibility to be a responsible AFRaKaN intellectual and work hard for the complete liberation of AfRaKa and the Neteru. In many ways, to accomplish this goal, I carefully studied AFRaKaN philosophy and history, intending to develop a new perspective that could be used to liberate the studies of AFRaKa from Caucasian hegemony. The development of this perspective was derived from the work and scholarship of many intellectuals in our community, such as Dr. Ben, Dr. Clarke, and many others whose efforts helped to recover and restore AFRaKa to her rightful place in history.

In this country, there has been a hidden agenda to deny and reject the fundamental principles of AfRAKaN culture, leading to a chaotic state. This also leads to sickness, which eats at society's very fabric. If not corrected, we could be confronted with an extinction-level event as the conflict between the world's peoples continues to rise. In this regard, not only must the feminine component be restored to her rightful place, but we must redefine and restore also the existential place of the masculine component. In this regard, developing an understanding of Science in addition to traditional AfRaKaN culture is very important not only for AfRaKaNs but for all beings. Religion has led to a state of confusion that doesn't exist outside of Eniyan.

This is why the advice my obi gave influenced my development because it shaped my identity. When I couldn't remember anything else, I remembered that, and this society decided to torture me. In many ways, although I am the omo of Iya, I also have Nna's energy and exist as a reflection of both. This is a lesson that my DNA confirms and that a wise group of awon arabinrin and arakunrin helped me to remember as part of my recovery. Although my memories of Nna are limited, it resonates within my identity and throughout my existence. In many ways, like Nna, I represent the struggles of our people to recover our understanding of pr ankh and return to AFRAKa.

This is why my goal is to continue my recovery and studies to increase my understanding of traditional AfRAKaN culture and reverse the spell by liberating my Neter so that I can be fully restored and serve as an Oluko for others. Of course, I realized there would be a lot of resistance to this, which made me realize that I should remain silent and instead put my thoughts into my writing. In many regards, this reflects the wisdom of my obi and especially the guidance I received from Nna.

But what I must also acknowledge is that this seemed to be the same message presented to me by other AfRaKaN Okunrin who served as my awon oluko like John B. Mullins, Dr. Joseph E. Lewis and the other great teachers that I had. They taught me the fundamental principles and concepts of AfRaKaN History and culture. Professors Lee Ingham and Dr. Jeff Crawford taught me Philosophy and the Principles of Reasoning and Logic.

At the same time, I also studied Computer Science at Central State University in Ohio. At Central State, I also had a few continental AfRaKaN teachers like Dr. Owindiwe, who taught Political Science and the fundamental principles of Pan AfRaKaNism. As a result, along with a group of enlightened awon arabinrin and arakunrin, we decided to start a local chapter of the OAAU. We became very active in local politics while most pursued plans to repatriate back to AfRaKa.

At this time, I had completed 22 awọn iyipo around Oorun and decided that I needed to extend my Education, so I applied for several graduate programs, including Howard University and Ohio University; However, Howard accepted me with a full scholarship, I never received the notification. So, in this regard, I enrolled at OU and a Master's in Philosophy and became very active in the AfRaKaN in American Student Union. But due, confrontations with Caucasians thinking over whether or not Ancient Kemet was AFRAKaN and that AfRaKaNs were capable of critical thinking caused numerous problems for me.

By surprise, I received a call from a Professor at Howard named Dr. Jeff Lambord, who asked why I rejected the

Full scholarship at Howard. I informed him that I had never received a notice, so I wasn't aware I had received a scholarship. At that time, Dr. Lambord asked me if I would still be interested, and I said yes, then asked when I could be there and replied Tomorrow. We ended the conversation, and I called home to Iya in Ohio and informed her about the great news, and asked Iya to join me on the trip to Howard.

CHAPTER III:
The influences of Education on shaping my existence

After graduating from Central State, I spent time teaching as a substitute in the Cincinnati Public School System. In that capacity, I would introduce the young students to AfRaKaN History and Politics and teach them the Principles of Reasoning to develop their critical thinking capacity. The most exciting aspect of this was the ability to reach out to young students and encourage them to connect with their AfRAKaN origins. Many students would ask questions about AfRaKaN, and most were explicitly interested in Ancient Kemet.

At that time, I was fascinated with Abyssinian due to the heavy indoctrination of Cacausianism in this country. Still, I knew that my heritage was primarily from "West" AfRAKa. My ebi living in Alabama, knew this well, which was incorporated into our identity without help from any religion. Iya Richeta was adamantly against faith, as were many of my relatives. However we did have Jehovah's Witnesses in our family, but they were often shunned and not accepted. In my regard, my ebi encouraged me to keep an open mind and to follow in the footsteps of Nna.

My teachers and my Education reinforced this sentiment. So, I tried to follow in the footsteps of my Nna carefully, not understanding that this society places a high premium on masculine energy only to create a high-level and productive workforce. This means that the AfRaKaN existence in America is driven by labor, and AFraKaNs are subliminally driven to be enslaved to the economy, except for a few wealthy AfRAKaNs here. Understanding this allowed me to appreciate Nna more because this was something Nna knew well, which motivated him to connect with AfRAKa. In my opinion, like many others, Nna keeps this to himself to avoid ridicule and condemnation.

When I was younger, I couldn't understand this due to the allusion and euphoria of freedom and fell into the "Black" trap, which is space cut out within Caucasianism for the converted AfRaKaNs. As I mentioned in The Book of Iya series, we have been dominated and subjugated to the Germanic Weltanshuung (worldview) and are slowly being defined out of existence. Because we are unaware of our history and historical enemies, we become casualties of the trap.

To follow the advice of Nna and my awon oluko, I studied at Central State under the tutelage of Dr. E. Lewis. I then also continued to teach while a Graduate of Howard University. That also allowed me to teach at Bowie State University and the University of the District of Columbia. My area of expertise was in Philosophy, where I taught the Principles of Reasoning, Logic, and basic Philosophy and began focusing on development.

This experience enhanced my understanding of AfRAKaN culture and gave me a sense of how the methodology could liberate AfRaKaNs. But at the same time, I started to train for IT certification and focused on Microcenter's MCSE for Engineers. This was due to the realization that modernity had been colonized and subjugated, and the only way out was to update the paradigm associated with AfRAKaNcentricity.

Teaching allowed me to test further and tweak my understanding of the method in many ways. To this extent, I started to teach the themes of traditional AfRaKan culture and history using the principles of AfRAKaNcentricity. This also gave me a greater understanding of what it meant to be AfRaKaN and allowed me to reject the Caucasian attempts to colonize my identity.

This is a war that can't be waged or fought with tanks and missiles because that arena is already dominated. On the other hand, we can fight to restore our consciousness and cultural identities to become fierce warriors just like our ancestors. In my regard, I traced my ancestors back to Nubia, Ardi and Lucy, so when we go to battle, we need Kandakes to fight with us side by side. This will allow us to restore and preserve our existential identity and reject the Caucasian Identity Spell.

Like my obi taught me, as awon okunrin, we stand in defense of our ebi, but to face this existential threat, our awon obinrin must also rise and stand with us. But at the same time, we have been dominated by masculine energy, only that no room is made for the allowance of feminine power. This makes Nna's advice so significant because I always try to recognize the natural balance in everything.

This also allows us to avoid Caucasian imperialism, which is about mind control and the refusal to pay reparations. This does the research of awon arakunrin like Garfield Reid, Bro Kofie Piesie and Ini-Herit Shawn Khalfani Phillips, Asar Imhotep, and Ankh so important as moving forward with the complete liberation and restoration of AfRaKaN concepts of existence. In this regard, while the ebi is being torn down, we aim to lift the ebi back up to be in a defense position. And so I stand on solid soil with Geb by my side with the energy needed to move forward in honor and balance. With that, I am reminded of the many awon iyipo spent in Ohio. After finishing my early years in Madison, Al, my ebi moved to Cincinnati, Ohio, where I was enrolled in the public indoctrination system.

Most of my Education was experienced in Ohio. From Brookhaven Elementary to the Princeton School System and back to CPS, where I finished Junior High and then High School at Robert A. Taft. I graduated from High school with honors, ranked number 5 in my class, and then enrolled at Central State University.

During that time, my egbon Chuck and I came up with the idea to start a Vending Company called Pride Industries. We put vending machines all around Cincinnati, including some at my High School and other places. Egbon Chuck also purchased an Ice Cream Parlor at the University of Cincinnati which we managed together with my egbon Anthony. Entrepreneurship is a theme that runs heavily in my ebi.

Richetta and Walter drove the foundation to motivate our family to be self-sufficient and independent. At the same, Aburo Lenny served as the Vice President of Micro Metal Finishing and would also give us excellent management advice. This would assist as I wrote up business plans and ensured that we keep with Local, State, and Federal business requirements. I began to learn everything I could to help manage our company. Learning to use software like Quickbooks and TurboTax was a great asset and allowed us to operate officially and adequately. Existence on the oko and working in the family store helped me to develop vital accounting and business skills. I understood that my ability and skills in Mathematics extended from all the experience I gained from living on our oko.

This also provided a source of motivation that was always derived from my obi and my ebi in general. In this regard, I credit my awon aburo because they had the same sense of entrepreneurship. Especially mi awon aburo Melvin, Nora, Joe, as well as mi aburo Lenny. One thing I can tell you about them is that they were awon Okunrin that loved and honored awon Obinrin. I remember sitting out at night, glazing at Sky, and chatting about history. Mi awon aburo would often pull out their rifles, shoot into Sky, and tell entertaining jokes.

CHAPTER IV:
Learning the fundamentals of (Pan) Afrakanisms at Central State and Howard

One of the benefits I derived from the coma and the experience I had with the insurance company is it allowed me to reflect on my existence and restore my identity. I also used the time to honor and defend my ebi. Given the challenges I faced, this was not an easy task. But who would have thought that it was illegal to AfRaKaN in America? With all the hate and everything perpetuated against Africa, it's hard to see any light at the end of any tunnel. Instead, the path for the AfRAKaN in America is filled with roadblocks, and this serves as a driving force straight into religion. Unfortunately, many of us remain caught in the trap, not realizing that all the Abrahamic faiths are indeed Caucasian.

These religions stem from Persia, Mesopotamia, and Canaan and are agents of Caucasian Imperialism. We failed to realize that what the Germans did was a design and develop an Identity spell heavily embodied in the Caucasian languages. Learning Yoruba allows me to redesign and develop my sense of identity. While I have recovered completely, as a Nna, I have done my best to defend my ebi. I have paid a severe price only because I refused to submit to existential imperialism.

When I started writing the Book of Iya series, much thought and made many false accusations against me, they even assumed I was attacking their leader. Instead, after escaping the Abyssinian mind control trap, I concluded that this society is full of elitists trying to escape global warming and have the masses enslaved who don't know it. Just because something sounds good doesn't mean it's true, and that's why I let go of religion so I could return to AfRAKaN existentially and honor my Iya and defend Nna. It hasn't been easy, but I find strength when I see my ebi smile.

Creative and free-thinking Continental AfRaKaNism is not allowed in the United States and is penalized at the highest cost. I could stay strong and push my recovery forward while this society stood by and watched. But thanks to the strength of my ebi and the of many others, we have survived. Currently, I would like to visit Nna because, due to my ailments, I haven't seen Omowale in years now, so I am thinking of driving to Madison, Al, to visit my ebi there. I will thank Nna and thank him for giving the advice that helped in my recovery so far. I'll let Nna know that I will never betray Iya and will always defend Nna.

When I started writing the Book of Iya series, much thought and made many false accusations against me, they even assumed I was attacking their leader. Instead, after escaping the Abyssinian mind control trap, I concluded that this society is full of elitists trying to escape global warming and have the masses enslaved who don't know it. Just because something sounds good doesn't mean it's true, and that's why I let go of religion so I could return to AfRAKaN existentially and honor my Iya and defend Nna. It hasn't been easy, but I find strength when I see my ebi smile. A creative and free-thinking continental AfRaKaN perspective is not allowed in the United States and is penalized at the highest cost. Regardless, I could stay strong and push my recovery forward due to the strength of my ebi and the assistance I received from many others.

Currently, my goal is to visit Nna because I haven't seen Omowale in years due to my ailments, so I am thinking of driving to Madison, Al, to visit my ebi there. I will thank Nna for providing the advice that has assisted my recovery. I'll let Nna know I will never betray Iya, and TaNehesi will always defend It Neter/Nna.

Bibliography

Books

Afrika, Llaila Melanin What Makes Black People Black! New York: Seaburn Publishing Group, 2009

AfRaKan Academy of Sciences. Workshop on Science and Technology Communication Networks in AfRaKa. Nairobi: AfRaKan Academy of Science, 1993

Balaam, David N., and Michael Veseth, eds. Introduction to International Political Econo New Jersey: Prentice Hall, 1996

Baradat, Leon P. Political Ideologies. New Jersey: Prentice Hall, 1994

Billet, Bret L. Investment behavior of Multinational Corporations in Developing Areas. New Brunswick: Transaction Publishers, 1991

Clough, Michael. Free at Last? New York: Council on Foreign Relations Press, 1992

Dalley, Stephanie. Myths from Mesopotamia. New York: Oxford University Press, 1992

Drew, Eileen P., and F. Gordon Foster, eds. Information Technology in Selected Countries. Tokyo: United Nations University, 1994

Dubois, W.E.B. The World and AfRaKa. New York: International Publishers, 1965

Fieldhouse, D.K. Black AfRaKa: 1940-1980. Boston: Unwin Hyman, 1986

Haggard, Stephan, and Robert R. Kaufman, eds. The Politics of Economic Adjustment. Princeton: Princeton University Press, 1992

Harbeson, John W., and Donald Rothchild, eds. Afiica in World Politics. Boulder: Westview Press, 1991

Leedy, Paul D. Practical <u>Research: Planning and Design</u>. 5th ed. New York: Macmillan Publishing Company, 1993

Men-Ib Iry-Maat, Wudjau. <u>A Beginner's</u> Introduction to Medew Netcher 2nd ed. USA Wudjau Men-Ib Iry-Maat, 2016

Moran, Theodore H. <u>Multinational Corporations.</u> Massachusetts: Lexington Books, 1985
National Research Council, Office of International Affairs, <u>Bridge Builders.</u> Washington: National Academy Press, 1996

Obenga, Theophile. <u>African Philosophy.</u> USA: Brawtley Press, 2015

Piesie, Kofi. <u>Beautiful Lessons About Kimoyo</u> USA: Same Tree Different Branch Publishing, 2021

Piesie, Kofi. <u>Spear Masters,</u> USA: Kofi. Piesie Research Team, 2021

Rodney, Walter. How <u>Europe Underdeveloped</u> AfRaKa. Washington: Howard University Press, 1974

Reid, Garfield. Misconceptions & Misinformation by the Black Hebrew Israelites Vol 1. USA: Garfield Reid, 2021

Rosenbloom, Richard. Technology and Information Transfer. Boston: Harvard University Press, 1970

Sandbrook, Richard. The Politics of AfRaKa's Economic Recovery. New York: Cambridge University Press, 1993

Segal, Ronald. Islam's Black Slaves. New York: Farrar, Straus, and Giroux, 2002

Shibre, Zewdie, and Abdulhamid Bedri, eds. Regional Development Problems in AfRaKa. Addis Ababa: Institute of Development Research, 1993

Slater, Robert O., Barry M. Schutz, and Steven R. Dorr, eds., Global Transformation and the Third World. Boulder: Lynne Rienner Publishers, 1992

Steindorff, George and Seele, Keith C. <u>When Egypt Ruled the East.</u> Chicago: The University of Chicago Press, 1942

Turabian, Kate. A <u>Manual for Writers.</u> 5th ed. Chicago: University of Chicago, 1987

Weiss, Thomas G., and Merl A. Kessler, eds. <u>Third-World Security in the Post-Cold War Era</u> Boulder: Lynne Reinner Publishers, 1991

Weston, Alan F. <u>Information Technology in a Democratic</u> Cambridge: Harvard University Press, 1971

Articles, Papers, and Public Documents

Da Costa, Peter. "AfRaKa-Communication: Internet - A Statist Model," Addis Ababa: International Press Service, September 10, 1996, National Telecommunications and Information Administration, "U.S. Goals and Objectives for the Information Society and Development Conference", prepared remarks of Vice President Al Gore, delivered via satellite to the Information Society and Development Conference in Midrand, South AfRaKa(May 13, 1996)

Semret, Nemo. "Unleashing AfRaKa's Potential: The Technological Reasons for Open and Competitive Cybercommunications", a paper delivered at The Second Annual Meeting of the AfRaKa Scientific Society, Washington, June 22, 1996

Burka, Lauren P. "A Hypertext History of Multi-User Dimensions." MUD History. 1993. http://www.utopia.com/talent/lpb/muddex/essay (2 Aug. 1996).

Fine Arts." Dictionary of Cultural Literacy. 2nd ed. Ed. E. D. Hirsch, Jr., Joseph F. Kett, and James Trefil. Boston: Houghton Mifflin. 1993. INSO Corp. America Online. Reference Desk/Dictionaries/Dictionary of Cultural Literacy (May 20, 1996

Tanehesi The Restorer Vol 1, 2, and 3

©SameTreeDifferent Branch Publishing

i

www.ingramcontent.com/pod-product-compliance
Lightning Source LLC
Chambersburg PA
CBHW041926090426
42743CB00020B/3457